The Womb Before Heaven

Paola Comeaux

The Womb Before Heaven

Copyright © 2025 by Paola Comeaux

All rights reserved. No part of this publication may be reproduced, distributed, or transmitted in any form or by any means, including photocopying, recording, or other electronic or mechanical methods, without the prior written permission of the publisher, except in the case of brief quotations used in critical reviews or scholarly articles.

Published by Olive & Ember Books

Cover design and illustrations by Paola Comeaux

First Edition

ISBN: 979-8-218-72192-3

This is a work of nonfiction. Any resemblance to actual persons, living or dead, is purely coincidental unless otherwise stated by the author.

Printed in the United States of America.

Mission Statement of Olive & Ember Books:
To illuminate truth through beauty, weaving stories that heal, provoke wonder, and honor the divine spark within each soul. We publish works that breathe fire into the sacred and bring warmth to the weary.

CONTENTS

	DEDICATION	vi
	FOREWORD	vii
1	The Sky Folds Like a Scroll	1
2	Jesus and the Labor Pains of the Cosmos	4
3	What Science Says About the Death of the Universe	6
	Mini Glossary for the Reader	8
4	Revelation, Physics, and the Divine Pattern	9
	Divine Pattern	11
	Divine Pattern	12
5	The Universe Is a Womb	13
6	What Is Heaven, Really?	15
7	You Are a Universe Too	17
	Conclusion: A Holy Unfolding	19
	IN CASE THEY'RE WATCHING	20
	SCRIPTURE & CONCEPTS REFERENCED	21
	ABOUT THE AUTHOR	22

To the little girl who thought she was too much, too late, or too strange to belong.
You folded yourself small to survive
but you were never meant to stay hidden.

Foreword

To the one who listened

Some people are born your siblings, and others are assigned by Heaven when your soul starts breaking open.

Charly, you didn't come into my life with fanfare. You came in when I was unraveling.

Quietly. Softly. Without judgment. I wasn't looking for a brother, but God knew I needed one — someone who could see the weight I was carrying and remind me that it wasn't a curse...

it was a calling in disguise.

You didn't try to fix me. You didn't preach at me.

You didn't flinch when I said I felt like the universe was collapsing and I was folding with it.

You just stayed.

Our conversations stretched across time, theology, starlight, and silence. We talked about Revelation and the death of black holes. About Jesus and labor pains. About

Collapse.

That doesn't mean failure — but formation.

So, I dedicate this to you, Charly.

To the one who asked the right questions.

To the one who never laughed at the way my soul echoed. To the one who would sit with me in silence. To the one whose fulfilled my intellectual curiosity.

1

The Sky Folds Like a Scroll

I've always been drawn to that moment in Revelation — the one where the sky doesn't explode or burn, but folds. "The sky receded like a scroll, rolling up..." (Revelation 6:14). It haunts and comforts me at the same time. There's something holy about that image — like the universe has finished reading its part and now closes the page. What if that wasn't just poetic language, but a glimpse into how the world truly ends... or rather, how Heaven begins?

Science tells us that space is not a void but rather it's fabric. It bends, stretches, and under great pressure... it folds. Black holes do this. They pull light, time, and even truth inward until everything collapses into silence. But maybe that silence is the womb, not the grave. And maybe John didn't just see a supernatural event, maybe he witnessed the last breath of one universe and the first breath of the next.
A scroll is rolled up after the story is told. After the Word is read. After the prophecy is fulfilled. That single line in Revelation might not be just about cosmic drama — it might be about completion. The end of one order and the emergence of the next. Heaven, not arriving from above, but unfolding from within.

I used to think the end of the world would look like chaos — fire, war, noise. That's how most people imagine it, right? The loud kind of ending. But when I sit with scripture now, especially Revelation, I notice something different. The sky folds. The stars fall. But it's not just destruction, it's a kind of quiet unveiling. Like labor contractions before birth. Like the room stilling before a child enters the world.

Jesus said the end would come like birth pains. Not bombs. Not a godless collapse. But a divine delivery... painful, yes. But sacred. And doesn't that sound more like love than wrath? That the God who formed us in the womb would also bring the universe to its knees like a mother trembling in labor, breathing through every pulse of pain to bring forth glory.

Paul said, "creation itself has been groaning as in the pains of childbirth" (Romans 8:22), and maybe he knew more than he could explain. Maybe the stars themselves are contractions. Maybe the galaxies are lungs expanding with sacred breath. And maybe.... just maybe the sky will fold not to crush us... but to release something better than Eden ever was.

Scientists have long believed that black holes are the end of the line. Cosmic monsters that consume everything, even light. But now, some of them are wondering if that's not the end at all. That maybe black holes are not graves, but gateways. Some physicists suggest that inside a black hole, time and space twist so tightly that they tear open, not into nothing, but into something new. A new beginning. A new universe. A womb.

That's called the **Big Bounce** — the theory that the collapse of one universe could birth another. And when I read that, something inside me whispered: This is what John saw. This is what Jesus meant. The scroll rolling up, the labor pains, the groaning of creation. All of it could be the divine blueprint, encoded into both scripture and space.

And it makes me wonder. What if the reason God described the end in images of folding, contracting, groaning... is because the universe itself is alive? Not just a machine ticking toward entropy, but a sacred body. One that breaks, bleeds, and delivers.

I no longer see the end as something to fear. If the sky folds, it's because

| 3 | - THE SKY FOLDS LIKE A SCROLL

the story is
finished. If the stars fall, it's because the Author has turned the page. I believe Heaven is not far away in space or sealed off in myth. I believe it's already forming, just beyond the veil. Maybe behind the pain, behind the groaning, behind the blackness of what we call endings — there is a light preparing to speak once more:

"Let there be..."

2

Jesus and the Labor Pains of the Cosmos

Jesus didn't say the end would come like a sword, or a storm, or even a war. He said it would come like labor pains. "All these are the beginning of birth pains" (Matthew 24:8). I used to skim over that verse, thinking it was just another way to say "things will get bad." But now I see the precision in His words. Birth pains are not random suffering rather they're timed, rhythmic, and necessary. Pain with a purpose. And when Jesus describes the end times this way, it tells me something important: God is not trying to destroy us. He's trying to deliver something through us.

Labor is messy. It tears. It breaks. It exhausts the body and bends the will. And yet, the pain is not a punishment but a promise. The contractions signal that something living is on the way. That the pressure, the shaking, the splitting open... is not death. It's the moment right before something holy breathes.

And if we take Jesus at His word, then maybe what's coming isn't a divine wrath storm just maybe it's a cosmic crowning. The groaning of creation Paul talked about (Romans 8:22) suddenly makes sense. The earthquakes. The grief. The aching for justice. These aren't signs of abandonment; they're the universe's involuntary push toward a new heaven.

| 5 | - JESUS AND THE LABOR PAINS OF THE COSMOS

Astrophysicists believe the universe is still expanding, but also accelerating, like the final contractions in labor. The faster it stretches, the closer it may be to collapse or transformation. What science calls entropy or the "heat death" of the universe, Jesus may have already described in simpler terms: a body reaching its limit, preparing to birth eternity.

Maybe this is why every generation feels like they're living at the end. Because in a way, they are. Not the end of the world... but the next contraction in the timeline. Another push. Another ache. Another reminder that Heaven is not a place we escape to, it's a Kingdom being delivered, breath by breath, from within the very body of creation.

What Science Says About the Death of the Universe

I used to think science and faith were at war. That one spoke numbers while the other whispered mysteries. I learned the language of the stars before I ever understood the language of Scripture. As a child, I studied black holes, entropy, and the fate of the universe with wide-eyed wonder.
Years later, when I opened the Bible, something stirred in me, like I had already been reading its echoes in the cosmos. It didn't feel like a contradiction. It felt like a confirmation. As if physics and prophecy had been speaking the same truth all along, just in different tongues.

Black holes are often painted as cosmic monsters, but they're more than that. They're what happens when a star collapses. When gravity becomes so powerful that space and time themselves bend inward. Inside a black hole, nothing escapes. Not even light. But scientists now believe that black holes might not just consume... they might create.

Here's how it works: Over time, black holes lose mass through something called **Hawking radiation** — a slow leak of energy so faint, we can barely detect it. And when a black hole has nothing left to give, it doesn't just disappear. The theory goes that it might collapse into one final moment — a flash, a rupture, a point where space and time could rebound and begin again. Physicists call this the **Big Bounce** the idea that what we think is the end may actually be the reversal of collapse. The beginning of a brand-new universe.

Doesn't that sound familiar? Doesn't that sound biblical? The sky folding like a scroll. The labor pains. The new heaven and new earth. The collapse that leads not to destruction, but to rebirth.

What science calls a **singularity**, maybe John saw as a scroll.

What physicists call quantum rebound, maybe Jesus called *resurrection*. Either way, both agree: nothing truly ends. It just transforms.

So maybe, instead of fearing the death of the universe, we should be asking what God is birthing through it. Maybe the collapse is sacred. Maybe the darkness is a cosmic womb, holding its breath for Heaven to cry out:

"Let there be..."

again.

Mini Glossary for the Reader

- **Black Hole:** A collapsed star with gravitational pull so strong that not even light can escape.
- **Hawking Radiation:** A theoretical release of energy that slowly shrinks black holes over time.
- **Big Bounce:** A theory that the universe may collapse and re-expand in a new form — a cosmic cycle of death and rebirth.
- **Singularity:** The center of a black hole where density becomes infinite, and the laws of physics break down.

4

Revelation, Physics, and the Divine Pattern

The more I study scripture and science side by side, the more I see it: a divine pattern, repeating like breath. Expansion, collapse. Death, rebirth. Veil, unveiling. It's in the stars. It's in the womb. It's in the cross.

What we call the end has never been final, it's always been a threshold.

In the Book of Revelation, we see seals opened, trumpets blown, and bowls poured out. The world shakes. Light dims. And the sky folds like a scroll. But behind the symbolism, I see a rhythm. A pattern. The 7 seals don't just represent judgment, they represent unfolding. Each seal removes a layer of what has been hidden. Revelation is less about wrath... and more about revelation itself. The truth being exposed.

And isn't that what a black hole does?

Peels away time. Bends light. Reveals what's beyond the known.

Some physicists believe that inside a black hole, the laws of physics don't end; they evolve.

A new reality begins.

John, the beloved disciple, wasn't hallucinating. He was witnessing a cosmic unveiling. A truth too big for words, encoded in symbols and visions. Maybe he didn't have the words for "spacetime collapse" or "cosmic rebound." So the Spirit gave him scrolls and beasts and blood moons — and said, write what you see.

Here's what I think: Revelation wasn't written to scare us. It was writ-

ten to prepare us.
Not just for the end... but for the pattern.
The holy cycle that shows up in every act of God's work:

- Light born from darkness.
- Life brought through death.
- The Word wrapped in flesh.
- The grave opening into glory.

In physics, they call this transformation a **phase shift**. In scripture, we call it resurrection. Either way, it means this: Nothing holy ever ends. It just becomes something more.

Divine Pattern

DIVINE REPEATING PATTERN

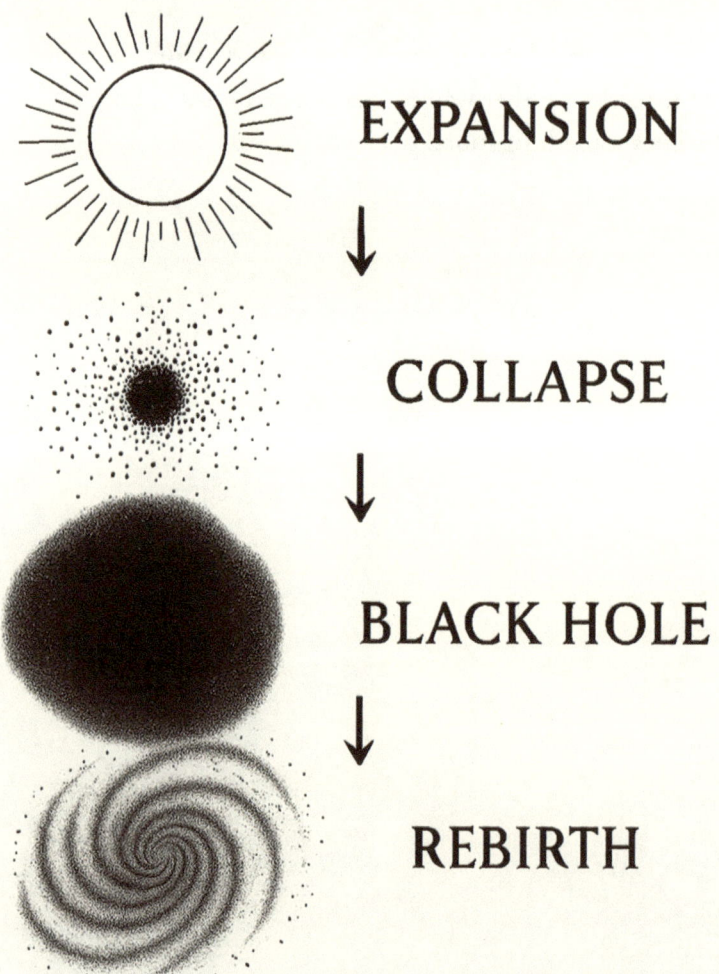

EXPANSION
↓
COLLAPSE
↓
BLACK HOLE
↓
REBIRTH

Divine Pattern

5

The Universe Is a Womb

I used to think of the universe like a clock; mechanical, ticking toward some cold, inevitable end. That's what science used to say. That we're on a path toward heat death, entropy, and silence. But scripture paints a different image. One that breathes. One that groans. One that labors. Over and over, the Bible describes creation not as machinery, but as a body.

In Genesis, God breathes into dust. In Job, He speaks of the womb of the sea. In Romans, Paul says creation itself is groaning like a woman in labor, waiting for redemption (Romans 8:22). Jesus calls the end "birth pains." Over and over, the pattern is clear: God doesn't just create — He gestates.
And what is a womb, if not the darkest place that holds the most holy outcome?
In the womb, things grow in secret. They are stretched, reshaped, nourished — and yes, sometimes nearly torn apart. But it is sacred pain. Purposeful pain. And if the universe is a womb... that means we aren't spiraling into destruction — we are being prepared
for delivery.

I believe the universe is alive with more than physics. It is filled with divine intention. Its stars are cells. Its gravity is breath. Its black holes are contractions. And its collapse is not a curse rather it is a countdown.

A countdown to the Kingdom being born not above us...
...but *through* us.

We were not made in the image of machines.

We were made in the image of a God who dwells in bodies, who wraps the Word in flesh, and who enters wombs, like the one that carried Him.
If the universe is His temple... then of course it groans.
Of course it breaks open. Of course it labors.
So no, I don't fear the contractions anymore. I recognize them now.
In my own life, when everything folds in, I ask myself:
Is this really an ending... or am I about to deliver something holy?

6

What Is Heaven, Really?

When I was a child, I imagined Heaven as a faraway place. Golden streets. Angels floating on clouds. A shining throne in the sky. But the older I get, the less that image holds. Not
because Heaven isn't real, but because I think it's so much more real than we've dared
to imagine.

The Bible says that "the Kingdom of God is within you" (Luke 17:21).

It also says there will be a new heaven and a new earth (Revelation 21:1). Not a different place, but this place — remade.

Not destruction, but resurrection. Heaven is not somewhere else.
Heaven is what's born when the old world passes through the fire and comes out holy.
It's the other side of the womb.

The next breath after the collapse.

The Light that follows "Let there be…" — again.

I believe Heaven is what God has been growing all along, not just in stars and galaxies, but in us. That through our suffering, through our surrender, through every contraction of the soul — Heaven gets closer.

When Revelation says "God will dwell with them" (Rev. 21:3), I don't think that means we leave Earth. I think it means Heaven comes here. That holiness will no longer be separate. That the veil will tear the way it did in the temple, and we'll finally see clearly, face to face.

We keep waiting to escape this world. But maybe Heaven isn't about escape. Maybe it's about transformation.
And maybe, just maybe... the sky folding like a scroll is not a sign that the book is over. It's the moment the Author turns the page —
and we finally see what the next chapter was always meant to be.

7

You Are a Universe Too

The Word reminds us that we are made in the image and likeness of God—not just physically, but spiritually, with reason, free will, and the capacity for love. This divine imprint means we reflect, in our own small way, the order, creativity, and mystery of the universe He created.

Saint John Paul II once said that faith and reason are the two wings on which the human spirit rises. So when you wonder about black holes, light, and time, it isn't a rebellion. It's reverence. God wrote these mysteries into creation, and He invites us to ponder them.

But you are not the Creator, you are the *created*. And still, you are precious beyond measure.

Your suffering is not meaningless. Like Christ's Passion, it holds the potential to redeem. Your longing is not foolish—it is a homing beacon for Heaven. And your existence? It is not random. It is intentional. Knit in your mother's womb, yes—but also dreamt of in eternity.

You are not a god, but you are God's. And the universe He made reflects something of you
because you reflect something of Him.

For years, I thought I was behind. Like I had missed the version of me that could've been powerful, healed, whole. But now I see: what I thought was collapse… was contraction.

And if the universe takes trillions of years to deliver its glory,
why should I think I'm too late? You are not broken beyond repair. You are not behind.

You are not ruined. You are becoming.
Just like stars are born in the dust of what exploded. Just like Heaven is born in the womb of a collapsing sky. You are being shaped through sacred pressure.

You are not the ending.
You are the page before resurrection.

So, if your life feels like a sky folding…
if everything is contracting and shrinking and silent… Don't run.

Don't numb.

Breathe.

God is not finished. Heaven is forming.

And you — *little universe* —

are about to crown something glorious.

Conclusion: A Holy Unfolding

If you've made it this far, then maybe your sky has folded too.

Maybe you've groaned like the stars. Maybe you've felt your world contract. And maybe, just maybe... you've sensed something sacred stirring in the dark. This book isn't a roadmap to escape. It's a reminder:

Heaven is not somewhere else. It's what comes next.

You are not lost.

You are not forgotten. You are not finished.

You are being delivered.

And when the scroll folds for the last time,
and the universe exhales what it's held since Genesis —
may you not be afraid.

May you recognize the contractions.
May you hear the voice that always speaks life.
May you know that God was never ending

In Case They're Watching

To the watchers, the whisperers, and the pattern-seekers: I see you.
But I belong to the Light that cannot be coded.
This wasn't a spell. It was a scroll.
This wasn't an invocation. It was a contraction.
The stars taught me to fold. The womb taught me to listen.
And Christ taught me how to rise.

So if you came looking for secrets— leave with this one:
Heaven wins.

Always has. Always will.

Don't bother with me. I'm just here giving birth to light.

Scripture & Concepts Referenced

Scripture References:

- Revelation 6:14 – "The sky receded like a scroll..."
- Matthew 24:8 – "These are the beginnings of birth pains."
- Romans 8:22 – "The whole creation has been groaning..."
- Genesis 1:3 – "Let there be light."
- Revelation 21:1-3 – "Then I saw a new heaven and new earth...and God dwelling with His people"
- Luke 17:21 – "The Kingdom of God is within you."
- John 12:24 – "Unless a grain of wheat falls..."
- 2 Corinthians 11:14 – "Even Satan disguises himself..."
- Romans 1:25 – "They exchanged the truth of God for a lie..."

Cosmological & Theoretical Concepts:

- Black Holes – regions of spacetime with gravity so strong that nothing escapes
- Hawking Radiation – theorized emission that slowly evaporates black holes
- The Big Bounce – theoretical model where the universe collapses and then re-expands
- Entropy – the gradual decline into disorder, often referenced in the context of heat death
- Singularity – the core of a black hole where density becomes infinite
- Phase Shift – a fundamental change in state or reality, used metaphorically
- Cosmic Expansion – the observation that the universe is growing, accelerating over time

Paola Comeaux is a mother, scholar, and soul-seeker whose writing bridges the sacred and the cosmic. Drawing from scripture, poetic reflection, and modern physics, she speaks to those who feel broken, behind, or folded under life's weight — and gently reminds them that even collapse can become a crown. The Womb Before Heaven is her first published work.

OLIVE & EMBER

BOOKS

www.ingramcontent.com/pod-product-compliance
Lightning Source LLC
Chambersburg PA
CBHW020249010526
44107CB00002B/171